Exploring

CHARLES C. WALCUTT
GLENN McCRACKEN

Consultants:
HELENE C. BROOME
GWEN FULWILER
MARETA H. VARNER

Lippincott Basic Reading B

HARPER & ROW, Publishers 1817 New York Philadelphia Hagerstown San Francisco London

Acknowledgments

Grateful acknowledgment is made for use of the following selection.

"Horns" by Austin Tighe. By permission of LAIDLAW BROTHERS, A Division of Doubleday & Company, Inc.

Photo Credits

Cover photo: Harvey Lloyd/Peter Arnold, Inc.; p. 70: Joel W. Rogers; p. 71: (left) George Hall/Woodfin Camp & Associates, (right) Joel W. Rogers; p. 72: (left) John Zoiner/Peter Arnold, Inc., (right) Joel W. Rogers; p. 73: Joel W. Rogers; p. 74: Walter H. Hodge/Peter Arnold, Inc.

Art Credits

Ruth and Allan Eitzen: pp. 46, 47, 57, 58, 114–117; Ethel Gold: pp. 6–8, 24, 25, 32–34, 60–63, 82–85, 119; Les Gray: p. 51; Cathy McCarthy: pp. 49, 65, 109, 120–122; Lucinda McQueen: pp. 41–44, 98–101; Heidi Palmer: pp. 52–55; Jan Palmer: pp. 27–30, 88–91; Dale Payson: pp. 1–4, 14–17, 104-107; Roland V. Schutts: pp. 36–39, 66–68, 103, 110–112; N. Jo Smith: pp. 5, 9, 12, 13, 18, 23, 26, 31, 35, 40, 45, 48, 50, 56, 59, 64, 69, 75, 87, 92, 97, 102, 108, 113, 118; James Watling: pp. 76–80, 93–96; Linda Boehm Weller: pp. 10, 11; Carol Wilde: p. 86.

Printed in the United States of America

ISBN 0-397-46630-7

CONTENTS

Fun in the Sun	1
Martin	6
A Summer Trip	10
A Trip to the Stars	14
Wags Gets Wet	19
Wes and Donna	24
Paw Prints on the Steps	27
Ann's Crown	32
Flat Sam	36
A Ball for Ben	41
The Little Sled	46
In the Dark	49
Pick Up a Stick	51
A Kiss for Mom	52
Twinkle and the Cow	57
The Skate Contest	60
Do I Dare . . . ?	65

Black Mane and Tim	66
How Lumber Is Made	70
Slick the Seal	76
The Lost Letter	82
An Ant	86
Ike's Pie	88
Dirk's Kite	93
Homer	98
A Little Elf	103
A Feast for Hope	104
Horns	109
The Sad Goat	110
A Ride in a Jet Plane	114
The Little Silver Robot	119
Valentine	122

Developmental Pages: ar—page 5; er—page 9; ed—pages 12, 13; wW—page 18; wa—page 23; wa, aw—page 26; ow—page 31; lL, ll—page 35; bB—page 40; le—page 45; kK—page 48; /k/ck—page 50; nk—page 56; /ā/ a_e—page 59; are—page 64; /ē/e, ee—page 69; /ē/ea—page 75; /ā/ai—page 81; /ī/i, i_e, ie—page 87; ir—page 92; /ō/o, o_e—page 97; or, ore—page 102; /ō/oa, oe—page 108; jJ—page 113; vV—page 118.

Fun in the Sun

It is hot in the sun.
Don naps on a rug.
Fran has a cup.
Fran digs in the sand.
It is fun.

Fran stands up.
Fran has sand in the cup.
Fran dumps the sand on Don.
Fran grins.

Don sits up.
Don is cross.
Don must get Fran.

Don has damp sand in his hand.
Fran runs from Don.
Run fast, Fran!

Fran stops.
Fran must rest.

Don runs up to Fran.
Don drops the sand on Fran.

Is Fran mad?
Fran is not mad.
Fran grins at Don.
It is fun in the sun.

ar

arm	art	car	cart
card	hard	harm	harp
far	farm	tar	tart
darn	dart	part	start
star	scar	scarf	garden

Martin

Martin must go to Dr. Hartman.
Mom starts the car.
Martin gets in.

It is not far.
Mom and Martin go in.

Dr. Hartman nods at Mom.
Martin has the mumps!

Martin must not get up.
It is hard for him to rest.

Pat sends Martin a card.
Pat did the art on the card.
It has a red star on it.

Grandma sent a card from the farm.
Grandma's card has a gift in it.

Martin's dad is in Barton.
Mom and Martin miss him.

A car stops.
It is Dad!

Dad and Martin rest on the grass.
Martin shows Dad his cards.

er

her	after	sister	camper
harder	farmer	under	partner
faster	duster	mister	hunter
runner	hammer	dinner	supper
dipper	hitter	summer	drummer

A Summer Trip

The Potters go on a trip in a camper.
It is hot in the camper.
The Potters stop to rest at a farm.

Dad and Mom get the picnic supper.
Pepper naps under Dad's hat.
Amanda fans her sister.

Amanda and Pepper run to the pond.
Amanda spots a frog in the grass.
Pepper darts after it.
The frog hops faster and faster.
It hops into the pond.
Pepper did not harm the frog.

Summer is fun for the Potters.

ed

added	ended	handed	mended
hinted	acted	dented	dusted
carted	parted	darted	started
rented	petted	nodded	rested
tested	hunted	spotted	drifted

ed

dumped	camped	crossed	missed
tripped	dripped	dropped	stopped
snapped	passed	hopped	pumped

farmed	harmed	grinned	hugged
dragged	fanned	hummed	tugged

A Trip to the Stars

Peg must rest.
Peg put a carpet on the grass.
Her dog ran up to her.
Peg and her dog sat on the carpet.
Peg and her dog had a nap.

The soft carpet started to go up.
The carpet passed the garden.
It drifted up past a nest.
Up. Up.
Faster and faster.

A car drifted past the carpet.
"Is it far to the stars?" said Peg.
"Not far," said the man in the car.

The carpet started to go fast.
"It is fun!" said Peg.
"It is fantastic!"
"Arf! Arf!" said her dog.

The carpet drifted past the nest.
It drifted past the garden.
The carpet dropped onto the grass.
Peg's dog tugged at her dress.

Peg sat up and grinned.
Her trip to the stars had ended.

w W

wet	went	west	western
win	wind	winner	winter
wag	wagon	wig	twig
twin	twist	swim	swimmer

Wags Gets Wet

The Wagners went on a picnic.
Pam had her red wagon.
Her dog, Wags, got into the wagon.
The wagon started to go fast.
Faster and faster it went.

Pam darted after the wagon.
Pam must stop it.
The wagon went faster and faster.
The wagon must not get to the pond.
Can Pam stop it?

Pam cannot stop the wagon.
The red wagon hits hard.
It is twisted and dented.

Wags did not stop.
Wags went into the pond.

Pam went into the pond after Wags.
Wags was not harmed.
Pam and Wags got wet.
Pam sat in the pond and rested.
Pam and Wags grinned.

wa

war	warn	wart	warp
ward	warden	warm	swarm
warmed	want	wanted	water

Wes and Donna

Wes wanted to swim.
Donna did not want to swim.
Wes said, "Get in the water."
Donna said, "It is not fun to get wet."
"A person has to get wet," said Wes.
"A person must get wet to swim."

Donna sat in the warm water.
Donna put on her swim cap.
"The cap is a wig," said Wes.
Wes dripped water on Donna's cap.
Wes hid under the water.

Donna dipped her hands in the water.
"It is fun to get wet.
It is fun to swim," said Donna.

wa

wand	wander	Wanda
swan	swat	swatter

aw

saw	sawed	paw
raw	draw	straw
dawn	fawn	pawn

Paw Prints on the Steps

Wanda and Ted are twins.
The twins ran to the apartment.
The twins stopped at the steps.

Ted saw mud on the steps.
Wanda saw paw prints.
Did a monster wander up the steps?
Did the monster go past Mr. Ward?

Wanda and Ted went in.
The paw prints went on up the steps.
Ted and Wanda ran past Mr. Ward.

Is the monster in the apartment?
The twins must warn Mom.
Wanda and Ted ran up the steps.

The twins stopped at the top.
The wet paw prints went on.

The monster is in the apartment!
Has the monster harmed Mom?
Can the monster harm us?

Wanda and Ted dart into the apartment.
In the apartment is a dog!
The dog is a gift from Mom.
Wanda and Ted hug her.

"It was not a monster," said the twins.
"It was MARTIN."

ow

how	now	cow	gown
down	town	frown	crown
crowd	crowded	drown	drowned

Ann's Crown

Ann put on Mom's tan dress.
Ann put on her mom's tan wig.
Ann put a crown on top of the wig.
Now the dress was a gown.
It was the gown of a star.

Ann sat down in front of a mirror.
Rags and Samson sat down.
"Wow!" said Ann. "I am not ten.
I am the star of the town."

"I must go to the tower," said Ann.
Ann went up the steps.
Rags and Samson went up the steps.

Ann got up on a pot.
Ann said, "Now I am in a tower.
Wow! I am the star of the town."
The crowd grinned at the star.

The pot tipped.
The crown went down.
The tan wig went down.
No tower.
No crowd.
No crown.
It was fun for Ann.

l L

let	letter	list	last
glad	flag	class	clam
law	lawn	claw	clown

ll

all	tall	call	small
will	hill	fell	tell

Flat Sam

Ten small clams sat in the sand.
The clams were under a log.
The water was warm and still.

Clams can dig in the sand.
Clams can dig fast.

Flat Sam saw the clams.
Flat Sam had a plan.
The plan was how to get the clams.

A clam called, "Dig, clams, dig!
Get into the sand.
Dig far down into the sand!"
The ten small clams dug fast.

Flat Sam swam in the warm water.
Flat Sam swam after the clams.
Flat Sam swam fast to the log.

The clams dug faster and faster.
Will Flat Sam get the clams?

Slap, slap went the water.
Flat Sam got up on the log.

The small clams hid from Flat Sam.
Flat Sam did not get the clams.
All the clams were down in the sand.

b B

bag	big	ball	bug
beg	bed	belt	bend
bump	barn	tub	scrub
rib	web	scab	grab
brown	better	number	lumber

A Ball for Ben

Small Ben begged for a ball.
His sister Anna got him a small ball.
Ben's cat saw the small ball.
The cat ran after it.
The cat hit it from paw to paw.
The ball went far under Ben's bed.

Ben said, "I want a big ball."
His sister got Ben a big ball.
Ben's cat saw the big ball.
The cat hit it under the tub.
Ben's cat is smart.
It can crawl under the tub.
The cat got the big ball.
Ben was upset.

"The cat hid the small ball," said Ben.
"It hid the small ball under the bed.
The cat hid the big ball.
It hid the big ball under the tub."
Ben begged Anna for a bigger ball.

Anna got a bigger ball for Ben.
Ben said, "It is the best ball of all."

Ben tossed the bigger ball.
His cat tossed the small ball.
The cat ran after the big ball.
Ben ran after his cat.
Ben and his cat had fun.

le

bubble	bottle	little	apple
cattle	saddle	middle	riddle
huddle	puddle	pebble	nibble

The Little Sled

It was the middle of winter.
Bob and Len sat on a little sled.
The sled started to go fast.
Down, down the big hill it went.

Bob held on to Len.
Len held on to the little sled.
"Stop the sled," begged Bob.
"How?" called Len.

The little sled hit a bad bump.
Bob and Len fell off.
But the little sled ran on.
It ran into a cattle barn.

The sled is bent and dented.
Bob and Len cannot sit on it.
Dad will mend the little sled.

Bob and Len are wet.
Bob and Len get warm in the barn.

k K

kid	kiss	kissed	kit
kitten	ask	asked	mask
milk	milked	park	parked
dark	darker	mark	marker
bark	kettle	kept	kidnap

In the Dark

In the dark,
In the park,
Kim met a kid in a mask.

Is it Ken?
Is it Kip?
Kim did not stop to ask.

ck

sack pack rack back

luck neck deck sick

pick stick sock lock

duck truck stuck clock

block black trick cracker

Pick Up a Stick

Pick up a stick
And hit the ball—
Hit it across the park.
The little black dog
Will run and run,
And get the ball back and bark!

A Kiss for Mom

Kip was sad.
Kip felt her mom did not want her.
It was dawn.
Kip got up.
Her backpack was under her bed.
Kip's warm socks were in the backpack.
Kip put in crackers, a cup and milk.

Kip stuffed apples in her pockets.
Kip bent down.
The backpack slid onto her back.
Kip crept down the steps.

The lock went **c l i c k**!
Kip was in luck.
Mom did not get up to stop her.

Kip sat down in the park.
A fawn was in the damp grass.
Kip fed the fawn her apples.

Ducks swam on the pond.
Kip fed the ducks her crackers.

A black kitten rubbed her leg.
Kip got the cup from her backpack.
The kitten lapped up the milk.

The sun was up.
Kip felt better.
Kip was not sad now.
The animals had helped her forget.

Kip's mom called her.
"Help me get waffles for us."
Kip ran back to kiss her mom.

nk

ink	sink	pink	rink
link	wink	sunk	hunk
drink	blink	skunk	trunk
bunk	bank	crank	tank
blank	blanket	sprinkle	twinkle

Twinkle and the Cow

The Banks have a farm.
Bob Banks has a brown cow.
The cow is kept in the barn.
Bob rubs her back and neck.
The cow blinks.

At dawn, Bob gets a bucket.
His brown cow stands in the straw.
Bob sits down and milks her.

Twinkle is the farm cat.
Twinkle begs for a drink.
Bob lets Twinkle drink warm milk.

a__e

can– cane	at– ate	Sam– same
pan– pane	rat– rate	mad– made
cap– cape	mat– mate	fat– fate
pal– pale	hat– hate	man– mane

The Skate Contest

The kids on Winter Lane want a skate contest.

The date is set for September 10.

Mr. Tate wants a safe contest.

All the skaters must put on helmets and pads.

Mr. Tate starts the contest.

"Get on the mark, get set, GO!"

Carmen starts to skate fast.
Wade is not far from her.
Carmen is in front of all the kids.
But Carmen has bad luck.
Carmen trips on a pebble and falls.
Now Wade is in front of her.

Winter Lane has a big hill.
The kids skate fast down the hill.

Nate wants to get in front of Wade.
Wade's left skate hits a stick.
But Wade gets up and skates harder.
Nate and Wade skate fast.
Arms and legs go faster and faster.

Wade is the winner!
Mr. Tate hands Wade the winner's cup.
Wade takes it and grins.
The cameras go click, click.

Mrs. Blake made a cake.
On top it said,
"The Best Skater on the Block."

are

dare	hare	rare	bare
care	fare	spare	scare
flare	glare	blare	hardware

car – care star – stare scar – scare

Do I Dare . . . ?

Do I dare skate and toss plates
to Nate at a fast rate?

Do I dare dress up as a clown
and go downtown?

Do I dare let Mark
park the ark in the dark?

Do I dare drink pink ink
from a sink and not blink?

Do I dare do flip flops
on a hilltop and not stop?

Black Mane and Tim

Black Mane is a tan mare.
Her mane is as black as tar.
Tim is a small, brown hare.

Tim and Black Mane are pals.
The mare and the hare nibble grass.

A big, black snake is in the grass.
It wiggles up to Tim.
The snake sits up and stares at him.
The snake scares Tim.

The snake stares at the mare.
The snake scares Black Mane.

Tim hops up on the mare's back.
Black Mane runs to the gate.
The gate is as tall as the mare.

"Help, Tim," calls Black Mane.
"The gate is locked!"
Tim helps Black Mane.
Black Mane runs to the barn.
The mare and the hare are safe.

Tim had helped Black Mane.
Black Mane had helped Tim.
The hare and the mare are pals.

e E

he	be	me	we

ee

see	bee	deep	keep
seed	feed	need	bleed
meet	feet	feel	week
sleep	green	street	fifteen

How Lumber Is Made

Pete's dad has a power saw.
He cuts down green trees.
The green trees are made into lumber.
Pete's dad is called a *faller*.

The faller makes little cuts in the tree trunk.

The faller saws a deep cut.

The tall tree starts to fall.

The fallers warn the men to keep back.

TIM-BER-R-RR!

Big claws pick up the logs.
The logs are put onto a log truck.
The big truck takes the logs to the sawmill.

Week after week, the logs bob up and down in the log pond.
A man feeds the logs into the sawmill on a belt.

Pete's grandpa is at the mill.

He runs a big saw.

Grandpa takes care not to get his hands cut.

The saw cuts deep into a wet log.

Grandpa saws the log into raw lumber.

Pete wants to be a logger.
Pete's dad and grandpa agree. . . .
Pete is free to be a logger
if he wants to be.

ea

eat	east	seat	meat
beat	heat	read	leap
tea	sea	seal	meal
mean	bean	leaf	dream
feast	least	treat	repeat

Slick the Seal

Slick is a big, black seal. Seals can swim fast in the sea. Seals can do stunts in the water.

Dean helps Slick do his stunts. Dean has bits of cod and clam in a bucket of water.

The kids sit in seats near the water.
The kids stare at Slick and Dean.
The little kids fear Slick.
But Slick is not mean.

Slick can do neat tricks for Dean.
Slick will do tricks for bits of clam and cod.
Clam and cod bits make a meal for a seal.

Dean hands a big ball to Slick.
Slick can go up and down a ladder on his flippers.
And the big ball will not fall!

The trick is hard for a seal.
But it is a treat for the kids.
The kids clap for Slick and Dean.
Slick is fed clam and cod bits.

Dean takes clam bits from his bucket.

He sticks a bit of clam on his beard.

Dean begs Slick to take the clam off his beard.

Slick leaps up and gets the clam bit.

He claps his flippers and begs for a treat.

Slick and Dean toss a ball into a basket.

Slick wins the game.

He leaps into the water for a treat.

Dean hands clam and cod bits to him.

Slick eats the treat.

He claps his flippers.

The kids clap.

The kids want to see Slick repeat his tricks.

ai

aim	mail	tail	fail
pail	sail	nail	hail
rain	brain	train	trail
paid	pain	paint	braid
drain	sprain	wait	waist

The Lost Letter

Miss Lake is paid to take the mail. Miss Lake is in the small town of Drain. Miss Lake has a lost letter and needs help.

"I will help," said Gail.

"Wait! I want to help," said Kate.

Kate held the mail cart. Gail held Miss Lake's cap.

It is hard for Miss Lake to take the mail fast. It is hard to see the numbers. The numbers need to be nailed up and painted black.

"Here is a lost letter," said Miss Lake. "The letter is from Spain. It is for Mr. Haines. His address is Fifteen Main Street here in town. But I did not see number fifteen on Main Street."

Gail and Kate grinned and ran down a trail.

"Wait for me," called Miss Lake.

The trail led to a trailer park. Mr. Haines was in trailer number fifteen.

"I was afraid the letter from Spain was lost," said Mr. Haines.

"I am glad we got the letter to the proper address," said Miss Lake. "Gail and Kate were a big help."

An Ant

An ant
Will pass
In a forest
Of grass
On a twig
Near a rain-
Puddle
Sea.
And a rock
Will be
A hill
Tiptop
And a fern
A monster
Tree.

Adele H. Seronde

i I

find	kind	mind	grind

i_e

line	nine	ride	hide
fire	wire	tire	hire

ie

pie	lie	tie	die
died	tied	tried	cried

Ike's Pie

Ike made up his mind to bake a pie. He planned to put the pie in a 4-H Club contest.

Ike made a lemon pie. He set the warm pie on the table. Ike went for a ride on his bike.

Iris and Dee are Ike's sisters.

"We can tease Ike," said Iris. "We can hide his pie."

Dee hid the lemon pie in back of the green rocker. Dee's kitten saw the pie and stepped on it. Now Ike's pie had a paw print on it.

Ike came in. His pie was not on the table. Ike tried to find it. Did Iris eat it? Did Dee hide it?

"The pie is here," said Iris. "We wanted to have a little fun."

"I will get the pie," said Dee.

Dee went behind the rocker. "The pie!" cried Dee. "The kitten stepped on it."

Dee and Iris felt sad. Ike's sisters helped him make a second pie. But Ike's pie did not win.

Ike was sad. His pie did not win. And Ike did not like his sisters to tease him.

"Forget it," Ike said to himself.

"Can we eat the pie now?" asked Dee.

Iris and Dee smiled. Ike's pie was the best.

ir

sir	fir	stir	bird
girl	dirt	skirt	flirt
first	twirl	swirl	swirled

Dirk's Kite

Dirk and the girls rested under a tree. "I am tired," said Dirk. "Let's not ride bikes."

"Let's sail kites," said Karen.

"But I still have to paint a cat on mine," said Irma.

"Help me sail mine," said Karen. "We can get the kites and meet at the park."

Dirk's kite had a black spider on it. Karen had painted a red bird on hers. The girls were first to get a kite in the air. Up and up went the red bird. The kite's tail swirled in the air. The kite seemed like a real bird.

Dirk forgot to make a tail for his kite. The kite went up but it did not go far.

Dirk made a tail and tied it on the kite. He wanted his kite to go up fast. Dirk held the line in his hand. He hopped on his bike and began to ride. His kite went up behind him.

"Stop! Stop!" cried Irma. "The kite will hit the wires!"

But Dirk did not hear her. He made his bike go faster and faster.

All of a sudden, the front tire hit a big rock. BAM! Dirk fell off and landed in the dirt. The girls ran to help him.

"The line snapped," cried Dirk. "Now the kite will hit the wires."

But Dirk was in luck. The kite began to fall down, down, down. It landed in a fir tree.

"The black spider must be bad luck for me," said Dirk. "Let's have a ball game."

o O

go	no	so	old
gold	told	cold	hold

o_e

cone	note	hole	pole
rode	rope	rose	nose
home	stone	smoke	alone

Homer

Homer was little and fat. He was brown and had black stripes. Homer had lots of legs, feet, arms and hands.

Homer liked to eat green plants. In the summer, he crawled up and down the tall grass.

Homer had a home in the rose garden. He slept alone in a little grass tunnel he had made himself. Homer liked to take a nap after his meals. He told Flo Mole he wanted to go to sleep. Homer tied a note on an old pole. The note said:

Homer went into his tunnel and fell asleep.

As Homer slept, it began to rain. It rained hard. Homer got wet and cold. He woke up. His nose was cold. All his socks were wet. All his feet were cold.

Homer tied a rope between blades of grass. He pinned his socks on the line.

The time came for Homer to make a nest. He made a little silk nest and went fast asleep. The nest kept Homer safe and warm in the cold wind.

In April, Homer woke up. He made a hole in his nest. He crawled up onto a twig. Homer sat in the warm sun. But he was not the same old Homer. Now he was a gold and black *mariposa*!

or

or for fork cork

horn born corn popcorn

acorn storm cord afford

ore

more core tore sore

store score wore before

A little elf
Sat all alone,
Alone in the lap
Of an acorn cap—
At home.

Adele H. Seronde

A Feast for Hope

Hope was ten now. Mom had started a big meal for her. The big meal was to be a feast for Hope. Mom needed more salt. Morris had a sore leg. So Mom asked Hope to ride the horse into town to get more salt.

It was cold. Hope wore a warm skirt and scarf.

On her ride to the store, Hope saw smoke. Black smoke was near the Mortons' cabin. "A fire," cried Hope. "I must ride back home and get help."

"Mom! Dad!" cried Hope. "The Mortons' cabin is on fire."

"I will get the wagon," said Dad.

Morris limped from the cabin. "Wait for me," he cried.

"No, Morris," his mom said. "The sore leg is still not healed. Wait here for us."

Mom, Dad and Hope rode off to help the Mortons.

Morris was left alone at home. "I cannot go to the fire," he said to himself. "But I can help get dinner for Hope. I will make the feast Mom started for her."

First, Morris limped to the garden to pick the corn.

"Now I can stir the apple cake and put it in to bake. I can crack nuts to put on top," he said.

Later, Morris put the best plates, forks and cups on the table. He added candles and a red rose for Hope.

At last, Dad, Mom and Hope came home. In the back of the wagon were the Mortons.

"It was a bad fire," Dad told Morris. "So we want to help all we can."

"The Mortons can eat dinner here," said Morris. "I made dinner for all of us."

"Hope saw the smoke and rode for help," said Mrs. Morton. "And now Morris has made a feast for Hope and for us. We are so glad to be here."

oa

oats	oak	boat	coat
loaf	soap	goat	toad
road	soak	croak	float
moan	groan	roast	boast

oe

hoe	toe	toes	goes

Horns

Little car's horn goes beep, beep.
Big car's horn goes BEEP, BEEP.
Little car, big car.
Little horn, big horn.
BEEP, BEEP, beep, beep.

Little truck's horn goes honk, honk.
Big truck's horn goes HONK,
HONK.
Little truck, big truck.
Little horn, big horn.
HONK, HONK, honk, honk.

Austin Tighe

The Sad Goat

An old goat ate grass and oats near the side of a road. He was a fat, old goat.

The old goat saw a brown toad hop down the road.

"Croak, croak," said the toad, as it went down the road.

"I want to hop like the toad," said the fat, old goat.

"But a goat cannot hop," said the oak tree.

"I can hop if I want to," said the old goat. "Do you want to see me? I will hop into the road now."

"Stop! Stop!" warned the oak tree.

But the fat, old goat did not stop.

The old goat tried to hop onto the road. But he did not hop far.

"Help," groaned the fat, old goat. "Help me get down," he moaned. "I will not boast again," he said to the oak tree. "I cannot hop like a toad. I hop like a fat, old goat."

j J

jet	job	jog	jam
jail	jump	junk	jar
jeep	joke	jaw	jacks
jacket	just	juggle	jigsaw
Joe	Jean	James	Joan

A Ride in a Jet Plane

Jack will go on a trip in a jet plane. It will be his first plane ride. He wants to see Jan in West Falls. Jack is eager to see her again.

A jet is a big plane and it goes fast. Jack has not seen a jet as big as the Star Fire.

It is time for Jack to get on. He will see the pilot as he goes to his seat. Jack must find seat 5-A.

Jack said "Hi" to James and sat down beside him.

"I have not been on a jet before," said Jack.

"Please put the seat belt on," said the hostess.

The jet started to go up and up and up. The cars on the roads got smaller and smaller.

Jack liked to press the little button on the arm rest. It made his seat go back.

James had a game. Jack was the winner the first time. James won a game. The last game was a tie.

"Here is dinner," said the hostess. Jack and James were glad to see her. It was fun to have dinner on a jet. And in no time at all, the pilot said it was time to land.

A little later the jet landed in West Falls. Jan and her dad came to meet Jack. Jack was glad to see Jan.

"How did you like the plane ride?" asked Jan.

"It was just fine," said Jack. "I had lots of fun on the jet."

v V

van	vest	vote	voter
save	wave	waved	cave
gave	five	dive	diver
drive	driver	alive	arrive
visit	over	silver	vacant

The Little Silver Robot

A golden beam lit up the dark forest. From the beam came a little silver robot. The robot blinked. His visit had started.

"Now," said the robot, "I will see if I can find a man."

The robot jerked his silver legs up and down until he came to a cave.

"I will go in," said the robot. "I must find a man."

In the cave, the robot saw an old coat and a stove. He spoke to the coat but the coat did not speak. He spoke to the stove. The stove did not speak. The little robot left.

Near the cave was a lake. A little duck swam and dived in the water. The duck soared in the air. It landed on the lake again.

The robot spoke to the duck. The duck spoke to him!

"Now I can go home," said the little silver robot.

Back at home, the little silver robot had to tell the big silver robots all he had seen.

"Tell me," said the big robot, "is a man like us?"

"No," said the little robot, "a man is not like us. I saw a man swim on a lake. I saw a man soar in the air. A man can even land on water."

"Fine job," said all the big silver robots.

V is for Valentine.
A is for Ann.
L is for love.
E is for eager.
N is for note.
T is for twinkle.
I is for *is it?*
N is for name.
E is for ED!

PHONICS CHART

Sound/Symbol Relationship Sequence

Starting Out, A

/a/aA (ant)
/n/nN (nest)
/r/rR (run)
/d/dD (dog)
/u/uU (up)
/m/mM (map)
/p/pP (pin)
/i/iI (in)
/s/sS (sun)
/o/oO (on)
/t/tT (ten)
/e/eE (egg)
/g/gG (game)
/k/cC (can)
/h/hH (hat)
/f/fF (fan)

Exploring, B

ar (art)
–er (farmer, runner)
–ed (ended, farmed, dropped)
/w/wW (win, warm, swan)
aw (saw)
ow (cow)
/l/lL, ll (let, all)
/b/bB (bed)
–le (apple)
/k/kK (kitten)
/k/ck (sack)
nk (bank)
/ā/a_e (made)
are (care)
/ē/e, ee (we, see)
/ē/ea (eat)
/ā/ai (rain)
/ī/i, i_e, ie (find, nine, pie)
ir (bird)
/ō/o, o_e (go, note)
or, ore (for, more)
/ō/oa, oe (coat, toe)
/j/jJ (jam)
/v/vV (vote)

Reaching Higher, C

sh (she)
ch, –tch (chin, catch)
th (then)
wh (what)
qu (queen)
xX (box)
yY (yes)
zZ (zip)
–ng (song)
–ing (jumping, sailing, winning, smiling)
–ed (handed, needed, stopped, waved)
–er (other, longer, baker, swimmer)
ir, ar, or, ur (girl, dollar, work, fur)
/ā/–ay (day)
/ī/–y (my)
/ē/–y, –ey, (happy, key)
/lē/–ly (safely)
soft c (cent, circus, fancy)
soft g (germ, giant, stingy)
–dge (edge)
–sion, –tion (admission, decision, motion)
short oo (book)
long oo (moon)
ow (slow)
ou (out, four, soup, your, young)
u, u_e (music, rule)
ue, ui (blue, suit)

continued on next page

Jumping Up, D

Formal review of sound/symbols in Texts, A,B,C
oi, oy (oil, boy)
ew, eau (few, beauty)
aw, au (saw, pause)
ph (photo)
gh (laugh)
ch (echo, machine)
silent w (write)
silent k (knit)

Rolling Along, E

silent b, l (comb, talk)
silent g, h, gh (sign, hour, right)
ea (head, great)
ear (earn, bear, heart)
/ē/ie, ei (field, ceiling)
/ā/ei, eigh, ey (vein, eight, they)
ough (rough, cough, bought, though, bough, through)
/i/y (myth)
/ī/uy, ui (buy, guide)
/i/ui (build)
/ə/ai (captain)
/e/ue (guess)
/əl/ile (missile)
silent t (listen)
silent n (autumn)

11 13 15 17 19 20 18 16 14 12 10